Table of Contents

A to Z Beekeeping
for Total Beginners

by Lisa Bond

Introduction

Welcome to this book. Inside you will find every-thing that you need to know about keeping bees. Now, before we go on, I must warn you, there will be some bee humor in here. There will be lots of puns too, I mean a lot! So if you do not wish to be entertained while learning, then I would suggest going out and buy-ing a boring book. This book is going to be fun.

Now down to business. Why would you want to keep bees? There are many reasons. Bees are dying out around the world, yet they help to produce up to one third of the world's food. We cannot let them die! By keeping bees, you are increasing the bee population and saving the world! Pretty cool, right.

Another great reason to keep bees is all the free honey that you will have access to. Soon you will be an expert, thanks to this book. That means that you will have honey coming out of your ears (not literally, that would be gross). Bees produce a lot of honey, as if by magic. Do it right and you could end up with a lot of great honey.

Now, not only will your honey be raw and organic (the best kinds), but it will also be local. Studies have shown that honey made from the pollen of local flowers can help to ease allergies and other ailments. The great part is, that even if it does nothing to cure or help any of those, you still have a great tasting sugar substitute.

You will also be able to smell the sweet aroma of honey wafting from your backyard (or wherever your beehive is) into your house. On a warm summer's day, this will be bliss. It will not only be our sense of smell which is tickled, our other senses will benefit too. Bees are friendly creatures and if you do not harm them then they will not harm you. It is a pleasure to see bees buzzing through your yard. Going from flower to flower and then back to the hive. The noise too. The gentle buzz of the honeybee brings any garden alive.

While we are talking about bringing your garden alive, we should talk about your garden being brought alive, literally. Bees will increase the pollination of your yard and the surrounding areas. If you grow fruits or vegetables then you will experience greater yields. If you have flowers, then those flowers will bloom like never before. Is there nothing that bees cannot do? Well, use doors thankfully. We like bees, but we do not want them all moving in to our house.

There are much more benefits that we could talk about, but for now, I will leave you with one thought planted in your mind. The honey that you harvest from your own bees will be the best honey that you have ever tasted in your life.

Where should your bees be?

There are many important things to consider when thinking about keeping bees, but the first thing you should think about is where you are going to house the bees. Before we look to the outside, we should look to the inside of your house. Who lives in the house? Start with these people. Do they want to keep bees with you? Are they going to be happy that there could be a beehive in your backyard or close by?

Kids are usually easy to get on board, but spouses my not be. A long conversation about the benefits of bees may help (or this book). Failing that you can always tell them, 'Don't worry, BEE happy'. That is a surefire way to convince someone that bees are okay (or to get a divorce).

Now that your family is on board, talk to the neighbors It may be legal to keep bees where you live, but an unhappy neighbor can make for a troublesome life. Talk with them about your plans and perhaps tempt them with the offer of some free honey. There can be a negative connotation surrounding bees, but that is most likely a smear campaign by the wasps.

We have talked about legalities, but it would not be a bad idea to double check the law surrounding backyard bee keeping, just to make sure that you are on the right side of the law. You may also find some great information from you local council. They may even offer information sessions or classes and can help you to properly plan your bee keeping adventure.

Even if you have talked to your family and neighbors (and they are happy for you to go ahead with your plan), there are still allergic reactions to take into account. Even if no one has had an allergic reaction to a bee sting, it does not mean that they are immune. If one of your family has never been stung, or even if they have only been stung once with no reaction, it may be a good idea to go to your family doctor to be tested for an allergic reaction. You could mention this to your neighbors too. Better safe than sorry.

You have talked with the people around, the hardest part is over, from here on in you only need to deal with bees. Scope out the area you have available for bees and begin to plan where you would like the hive to go. You have a couple of options with placement and there are some things to thin about.

You can have the beehive out in the open, or you could have it hidden from sight. If everyone is happy with the bee plan (or plan B) then you could have the hive out in the open. If your neighbors are unsure, or if you have children (or adults) who do not want to see the bees, you could think about coverage.

You may already have some trees or bushes in your yard. Placing the hive behind these can take it form the line of sight and the bees will not bee seen as much in your yard. If you do not have these available, then you could plant something or construct some sort of visual barrier. A screen or line of sunflowers provide a good way to hide the bees and bee hive.

If you do construct, or have, a barrier between you and your hive, you will notice that the bees fly out of the hive and upwards. They will also fly to the hive from a height. This will make encounters between bees and people less likely, if that is something that you are worried about.

We have considered the people in our placement of the hive, now it is time to consider the bees. So what should we be thinking about?

Bees like to rise with the sun, much as we do (unless it is the weekend, but bees have no concept of a work week). When placing your hive, you should be looking to have it in a place where the early morning sun will hit it. This will get the bees up early and out of the hive to forage for pollen. This also means that bees will not be leaving the hive in the afternoon or evening, which will likely be a busier time in your yard. Be careful with the sun though. If you live in an extremely warm climate, you should try and have some shade for the hives, so that when the sun has peaked in the sky, the hive is not hit with the immense heat that can overheat the hive. If the hive does overheat, then the bees will spend a lot of time trying to regulate the temperature of the hive instead of making honey.

It would be in your best interests to have a stand for your hive (which we will talk about later). This helps to protect from moisture. If a hive is placed on the ground, and made of wood, it will eventually absorb the moisture from the ground and damage can appear. As well as moisture, you should be thinking about the other elements. In places where the temperatures grow very cold in the winter, having the elevation will hep to protect the hive from the cold. The air can also circulate around the hive, but wind should be avoided if it can be. Bees are small and fragile creatures, so some protection from strong winds would be appreciated.

When thinking about the placement of the hive, you should consider how often you are going to be accessing it and how easy it will be to collect the honey come harvest season. You do not want to reap the rewards of your labour, only to have to carry seventeen pails of honey over a mountain, across a river and under a troll bridge. Make sure that you have easy access when the time comes.

The bees need easy access to. When they are flying in and out of the hive, they do not want to be disturbed constantly by people. If you place the hive with entrance facing away from any high foot traffic areas then the encounter between person and bee will be lessened. You can also place the hive with a slight decline towards the entrance, to help rainwater to run out instead of collecting.

Bees need water, so be sure to have the hive placed near to a water source. A fresh body of water with some organic matter in it is best. A pond, stream or river is ideal, but you could have a bird bath or some other container of water close by, just in case.

If it comes time to place your hive and you find that there is no suitable place for your hive in your yard, then there are alternatives that you can consider. You may have a family member or friend who would let you use their yard or land space. You could also find local communal gardens and talk to the people who have gardens there. They may welcome a hive if it would help with the pollination of their flowers, fruits and vegetables.

Get creative with your beekeeping. If you have no yard space then you could always put a hive on your roof (if it would be allowed and if there was easy access). It does not need to be a house roof, it could be the roof of a garage or shed.

How do bees form a hive? Do they wing it?

Before we talk about the actual hive, it would be good to have an idea as to how bees form a hive out in the wild. With everything organic, if we can mimic nature, we can get the best out of our animals, vegetables and more.

In the wild, bees create hives to have a place to live during the winter months and to have food to eat. The bees will find some sort of container for their hive, somewhere which has a lot of the properties we have already talked about (sun, wind, water, etc). This is often some sort of crevice or hollow. It could be the inside of a tree or a crevice in a rock.

When they have found a place for the hive, they will begin by constructing hexagonal tubes. These tubes take less material to make then other shapes. They also store more honey than most other shapes. They are also stronger than most other shapes. Bees are clever!.

To make the hexagonal tubes the bees will create beeswax (a great material which we will also talk about later), by chewing wax until it softens. They will then work together to construct these tubes. They come together to regulate the temperature of the hive, to be able to manipulate the wax as it softens and hardens.

Bees work their whole lives, which are all too short. Soon after they are born a wax-producing gland is formed in the abdomen. This is when the bee can start to help in the production of the hexagonal structures.

Once the structure is built, the bees are freed up to go out and forage for nectar. They find this in flowers. The nectar is stored in their pollen pouch and re-acts with an enzyme in there, this is the beginning of it turning into honey. When the bee returns to the hive, it transfers this changed nectar to another bee. From it's tongue to another bee's tongue. A gentle sweet kiss.

During this interchange, the liquid from the nectar is evaporating. The nectar is becoming honey. The bee will now be able to do some different things with this honey. The sugar within the honey is transformed into wax, which is then emitted from the pores of the bee. This is the wax which they chew to make the interior of the hive.

The remainder of the honey will be transferred to the hexagons for storage. Nectar and pollen will also be stored here too. The hexagons will also house the bee larvae until they are old enough to go out into the world.

When we look at what a bee can do and build, it is astonishing. Bees are marvelous creatures and by keeping bees, we are able to prolong their existence and increase their chances of survival. In return, we gain a lot of honey, honeycomb, beeswax and more.

Hives

Stand

It would be ludicrous, of us, to try and build a hive in the same way as bees do, but we can try our best and hope that they appreciate our efforts. Now do not buy or build a hive just yet, before you do that we need to talk a little about a hive stand.

Bees do not build their hive directly on the ground and neither should we. Some elevation can really help our bees and the hive. We talked a little about the need for air circulation, protection from winds, and temperature control. Before we construct a stand there a few more things to think about.

The ground beneath you hive should be level or have a slight decline towards the direction the door will be facing. If the land has a slight decline, then you do not need to worry about the stand or hive having a slight decline. Your ground should be clear. You do not want plants, weeds or grass growing up around your hive, nor do you want to have to worry about maintaining whatever is below the hive. Concrete, gravel or sand will work just fine, though it is completely up to you as to what will be under your stand.

Being able to view what is going on under, and around, your hive can be a great benefit to you. If you can watch the bees coming and going you can get an idea of the health of your bees. Being able to see the ground below will allow you to see how many of your bees are dying. Bee deaths are just a part of bee lives, you may have to hold a lot of tiny funerals.

Check what predators visit your yard and use that knowledge to decide on the type of stand and how high it should be. You do not want to lose bees, or honey, to some stinking animal.

When you are constructing your stand, you want to think about the hive on top of it. How high is your hive going to stand? Are you able to fully access the hive after to is on the stand? Is the hive too high? Or is it too low and you have to bend down too much for access?

Now that you have decided on the height of your stand, you can go out and buy one, or if you are handy you can construct one. There are many options for beehive stands on the market. Shop around and find the one that is right for you.

If you are constructing, try to build the stand into your yard. Perhaps you already have something that you can use. A hive can be placed on top of a table if the hive is not too high. You could construct a simple stand or box from two by fours. The stand could be made from metal or it could be a couple of cinder blocks placed together. There really is no limit as to what you can build your stand from. As long as you follow some of the guidelines above, you will be fine.

The Hive

The most common way for you to get your hands on a beehive is to buy one. Quite often you can buy a beehive kit, which will come with all the equipment you need to realize your beekeeping dreams.

If you search online, you can find many places form where you can buy a hive. These its will have all you need to know about putting your hive together and setting everything up. A better alternative, though, would be to buy the hive from a local supplier or bee enthusiast. If you buy local then you know that you will be getting a hive which will suit your needs. If you buy form a local supplier or enthusiast, you will also be gaining access to their knowledge.

When you buy a beehive, the most expensive option will be the hive which comes fully assembled. You can save some money by buying a hive which comes in parts and requires you to assemble them. You can save even more money by building the hive yourself. When buying a hive you should always try and buy from an esteemed seller with great feedback. A cheap hive that is poorly constructed will make a big difference to your bees.

If you built your own stand and it did not work out as expected, then I would suggest you go out and buy a hive, you do not want to get it wrong. If you are very handy, then you could build your own hive, though I would still recommend that you connect with the bee community in your location.

When building your own hive, there are still parts which will need to be bought separately. You will need to purchase the frames, which are the parts where the bees will build the hexagonal structures. The outer lid is something which is best bought if you are not confident in building one. You will also need to buy an excluder too. This part stops the queen bee from moving up into the parts where your honey production will take place.

Before you begin to plan and build your box, you should purchase the above items. The dimensions of the purchases will determine the dimensions of your hive. For the best hive, you will want to construct it from pine (untreated) or cedar.

The first thing to build is the bottom of your hive. Everything else will rest on top of this. This bottom board can be a solid piece wood, or it can be a screened piece of wood to allow ventilation. The board will be up to half an inch thick, but make sure that it is strong enough to house the hive on top of.

The next, and possibly the most important part, is the deep super. A super is a box where the bees will build a hive. The deep super will house the queen and the other supers will be where you get your honey from. You can have one or two deep supers per hive, but for our most basic hive, we are going to build just one. The deep super is basically a large box with no top or bottom, so four connected sides.

The dimension of the honey super, will be determined by the dimensions of the frames, which will sit inside your super. As the name suggests, a deep super, will be deeper than the other supers. You want to have enough room in each super to house eight frames. The deep super will house deep frames.

You can use tongue and groove joints to connect the sides, or butt the wood against each other and use nails to hold the wood together. You can use glue too if it needs it. Inside the deep super, you will attach two small strips of wood, one on either side, for the frames to hang on.

The last thing to do with your super is to make a small cut on the bottom of it (this can be done prior to assembly too). You want to cut a small strip from one side to act as the entrance. This strip should be big enough to let the bees in and out, but not so big to let predators or other bugs in.

Now that you have your deep super, you can place it on top of the bottom board. On top of the deep super, you can place the excluder. This piece has holes which are big enough for worker bees to fit through, but not big enough for the queen bee. This stops the queen bee laying eggs in the honey supers. You could manufacture this piece yourself, but the intricacy of the piece is better left to a professional.

Once you have your excluder you can construct your honey supers (of course, you can construct these beforehand, but I am giving the instructions in order of assembly). The honey supers will have the same length and width as the deep super, will not be as tall. The height will be determined by the height of the frames.

The honey supers are constructed in the same way as the deep super, just not as deep. Four sides held together, with two strips of wood to hold the frames. You do not need to add an entrance to the honey supers. The worker bees will be able to access them from the deep super. These boxes will give you your honey colored gold (honey).

Once you have your honey supers built (I recommend building two), they can be stacked on top of the deep super and excluder. We are almost done. There are only a few more steps to go, but the hard work is behind us.

You can build a cover if you know how to, but buying cover/covers is the safest option. You usually have two covers. One is an inner cover that fits inside the top super, this cover has a small hole in it to allow bees in and out. The outer cover is made of metal and protects the hive against adverse conditions.

Now, before final assembly, you can paint the box. This will help to protect your box from the elements and can help your hive to fit into a color scheme. This step is optional,. If you are going to paint your box, make sure that you use outdoor paint, it is non toxic and you do not paint the insides. The bees prefer to do their own interior decorating. Now that you have assembled each part and painted (if you chose to), you can assemble the entire box.

Assembly for the box is as follows:
- Hive stand
- Bottom board
- Honey super
- Frames
- Excluder
- Honey super/supers
- Frames
- Inside cover

- Outer cover

And there you have it. Make sure to take your individual pieces to where the hive is going to reside and assemble there. Now that you have your hive, it is time to get some bees. Woah! Slow down there. Before we get some bees, we should make sure we have all the necessary equipment.

Equipment

You may have one hive or you may have a hundred. It does not matter how many bees you are caring for, you will need the exact same equipment. You will see from the descriptions that some of the equipment is very necessary and some of it can be replaced by some items that you may already have around the house. The equipment I will describe is manufactured with beekeeping in mind. Replacement items will work, but they may not work as well.

Bee Suit

I am sure that if you imagine bees or beekeeping, your mind goes straight to the big white suits worn by beekeepers, the ones with the large mesh helmet. This piece of equipment is a must, mostly. You do not want to be stung and bees do not want to sting you.

While the entire suit is not a must, the veil most definitely is. It is designed to be ventilated and stop bees from getting in and around your head. If you do not want to go for the whole suit, then it is recommended that you wear clothing that covers all of your skin. You should also choose clothing which is thick and durable, like denim.

You should also wear appropriate footwear. Regular shoes and sneakers will be fine, though boots would be recommended. Bee suits can get hot inside, even the bee veil can get hot and chances are you will be working outside during the summer months. If you have the money and want to wholeheartedly move into beekeeping, you can opt for a ventilated suit. These suits are a little more expensive, but offer more ventilation to keep you cool. All in all, you are not going to be wearing the suit for a prolonged period of time, so if you are fine with the heat, then a regular suit or veil will be fine. You can always take a break or work when the day is cooler.

If you are making a quick visit to your hive, then you can always wear a hooded jacket, limiting the amount of skin exposure. If you are careful and do not antagonize the bees you should be fine.

Gloves are something which is often gone without. The advantage to going gloveless is that you will have more dexterity in your hands to work with the hive and bees. The disadvantage us that you will have a greater chance of being stung. If you buy gloves you will guarantee that they will be thick and durable enough to prevent bee stings. If you want to use your own or buy some regular gloves then you are open to the possibility of being stung. Thick rubber gloves should be fine, but there is always a chance of a sting.

Smoker

The smoker is an iconic piece of beekeeping equipment and is one of the most necessary too. When you go to collect honey from your bees, or if you go to just check on your bees, you are invading their environment. This can send them into a panic. Anything which is in a state of panic is not good. This can affect the health of the bee and can enrage them. An enraged bee is one which is more likely to sting.

To stop this from happening we use the smoker. Honeybees have a pheromone alarm system which is triggered by danger. One bee can alert the entire hive. When you use your smoker and blast some smoke around the hive, it acts as a barrier to these pheromones. The bees are unable to communicate while the smoke is in the air. They become calmer and less likely to be agitated and sting. With the bees in this state, you can go about the tasks you came to do.

The smoker is a simple device. It is a metal container, with a small chimney. It will come with a handle so that it can be held. It will also come with a small pair of bellows attached. When lit, the smoker will emit a small amount of smoke. The bellows can be used to push out more smoke when needed.

Hive Tool

This is a tool which you can do without, but life is so much easier with it. It is a small, flat piece of metal which tapers on each end. One of the ends will be sharp and the other will be curved. They come brightly colored and are one of the cheapest beekeeping tools.

They provide a lot of flexibility when it comes to the tasks they can do. The frames inside the honey supers can become stuck. Resin from the wood can build up and stick the frame to the super. Honey and wax are sticky substances too. The hive tool can be used to pry the frames from the super.

There are many other things that it can do. It can be used as a scraper, either to scare away the tree resin or to scrape the coating from the honey. It can be used to cut apart honeycomb. It can be used to squash unwelcome bugs. There are many ways it can be used.

Bee Brush

Bees are very vain creatures and will not venture outside to collect nectar until their hair is perfectly combed. I am joking, of course, there are no studies which show this to be true. A bee brush is a soft bristled brush which is used to remove best from any area in which you need to work. You can sue it to brush them from the outside of your hive, from the frames, or even from your clothing.

The bee brush is not an essential piece of the beekeepers kit, so having it is entirely up to you. Some people find it valuable to have with them, while others are convinced that it can anger or even harm the bees. The brush is cheap, so even if you find that you are not using it after you have bought it, you have not lost out on much.

Extracting Equipment

You have bees to produce honey, I hope, so there will come a time when you want to extract the honey from the hive. You will need some equipment to do this. We will talk more in-depth about honey extraction later, but there is a machine you can buy which will extract the honey through centrifugal force. The honey is then run through a strainer and into a resting chamber, where it sits before bottling. These pieces of equipment can set you back a little, but the equipment can also be rented from other beekeepers or bee keeping communities. There are also ways to extract the honey without this equipment and we will talk later about the benefits of both.

A Hive Full Of Bees

Choosing and Buying

Now that you have some knowledge of beekeeping, the location, the hive and some equipment, you need some bees. You could place an ad on Craigslist advertising your space and offering free rentals for honeybees or you could place a 'for rent' sign on the hive. The other option is to go out and buy some bees (or get some for free from a friend).

You will need to decide which type of bee you would like. There are three main types of bee which you can use. Most other bees are variants of these three, though there are other varieties out there. The three we are going to talk about here are good for beginners and are available in most places.Your options are:

- **Italian Bee**. These bees are the most common bee available and are usually bright yellow in color They are a very gentle bee. They are friendly and easy to manage. They also produce a good amount of honey.
- **Carniolan Bee**. These bees are darker in color than the Italian bees. Like the Italians, these bees are very gentle, but they take a little more effort in caring for them. The multiply quickly, so you will have more bees buzzing around your garden. They are very good for climates where there are hard winters.
- **Russian Bees**. These bees come in a variety of colors, from bright yellows to dark blacks. Their nature matches their color They can be very erratic.

On some days you will find them swarming together, other days they will not. They take their time
starting their hive, but when they do start they build
it extremely fast. They can take a lot more effort to
keep, but they are a hardier bee and less susceptible to disease.

The type of bee you choose may come down to
the environment it is going to be housed in or it may
come from personal preference. If you are going to
have more than one hive, then you can have more than
one kind of bee.

One of the more trickier parts of beekeeping is
finding a place to buy your bees. Most Walmart locations do not stock bees. I should clear that up. NO Walmart locations stock bees. That would make for a good
cartoon though.

When it comes tie to buy bees you have two options. The best and easiest would be to contact a local
beekeeper or community and talk to them. They may
be able to sell you some bees. They may want to give
you some bees. If you cannot get any bees from them,
then they will at least be able to point you in the right direction.

The other option is to look for a seller online. You
can often find ads on Craigslist, or other local sites, for
bees. It is possible that there will be a bee community
near you or a seller within driving distance.

Bee installation

Once you have decided on the type of bee and
have gone ahead and bought them, it is time to install
the bees in your hive. I just want to say, that if you have
made it this far in your bee journey, you deserve con-

gratulations. You have done well. You are about to create a bee community and help bees to flourish.

If you have someone around you who can help
you with the installation of bees in our hive, you will be
more confident when it comes to housing them. If you
are in this alone, do not worry, you can do this. It may
seem daunting and scary when you get to this stage,
but the installation of bees in your hive is pretty simple.
To make it easier for the bees have some sugar water
(one part sugar to one part water) and a spray bottle.
The spray bottle will be used to spray the sugar water
into the hive so make sure that it is a new bottle.

When your bees are delivered, or when you have
picked them up, make sure to keep the package shad-
ed, away from harm and properly ventilated. If you are
traveling a long distance, make sure to secure the
package so that it does not fall and to prevent any damage.

The bees will come in a small box or cage which
will have food inside for them. They can last for a few
days in the box, but the quicker you can get the bees
into a hive, the better. Spraying them with the sugar
water will keep them going for longer of they have to be
inside the box for a day or two. Spraying them with sugar water will also make them happier and less likely to
sting when it comes time to transfer them.

The cages that the bees come in will vary from
supplier to supplier. They may come in a cage which
needs a screwdriver to open. They may come in a cage
with a latch which can be opened by hand. It does not
really matter how they come as it will be easy to get
them out.

Try to transfer your bees when it is warm. Choose
a time of day when the sun is shining and the temperature is not too hot or cold. Have all the equipment you
will need, by your hive. Open your hive and spray some
of the sugar water onto the frames. This will keep the

bees fed and inside of the hive until they are acquaint-
ed with their new home.

Spray the bees one last time with the sugar water
to calm them and ready them to be placed in the hive.
Use whatever tools are needed to open the box, or
cage, containing the bees. Do not worry about bees flying at you, this will not happen. One or two may fly out,
but they will not all leave the box initially.

Stay confident and calm. The bees will sense this
in you and it will help to keep them calm too. When the
bees are exposed to the sunlight and the air, they may
begin to buzz more. This is perfectly normal. You can
spray them with more sugar water to calm them, and
you.

Once you have the main container ready to open,
hold it over the space in the hive, vacated by the
frames you removed. Open the container and tip it into
the hive, shaking it gently to help the bees move from
container to hive. If you have a bee brush, you can gently brush them out of the container.

The bees will be attracted to the sugar water that
you sprayed into the hive earlier. They will not have had
ready access to food, so the sugar water will be welcome. The queen bee (and sometimes a few helper
bees) will come in her own container. This container will
have a stopper at one end and a candy like substance
at the other.

The bees that are now inside the hive are not
used to this queen. They have no queen bee yet. It will
take time for them to become used to her and it will
take time for her to take her place as the queen. This
will take a little time, but this container will help.

If there is a cover over the candy-like substance,
remove it, but do not remove the candy substance. The
bees are ready for a queen. The queen is ready for her
bees. Place the container in the hive. The bees will be
able to see and smell the queen. Over a short period of
time, they will become used to her. They will slowly eat

through the and like substance at the end of her small container and when they have done so, the queen will emerge.

You can place this small container between two of the frames, but make sure to replace all of the frames once the bees are inside so that they can start building their hive within. Place the covers on top, you do not need to replace the honey supers yet. By leaving only the deep super, it gives them time to build a home for the queen.

Once they have filled 75% of the frames, you can place the next super into the hive. Repeat when this has been achieved in the new super until you have all your supers in place. The bees will stay mainly inside the hive for the first little while, but they will leave to use the bathroom. Since we did not build a tiny bee bathroom in the hive, they need a place to go. Bees are very clean. You would not excrete your waste in your house if there was no bathroom, would you?

Once you have closed the hive, you can retreat to your house (or yard), grab a beer, glass of wine, coffee or water. Hold it up in a toast to your new beehive and celebrate what you have just done. Congratulations you now have a fully functioning beehive. You should be very proud of yourself. You are now a beekeeper. You are saving the world, one bee at a time.

How To Care For Bees

Bees are very self-sufficient creatures, but they will need some care and attention. There are many things that you can do to help your bees as they live and grow in your yard.

What to focus on

Continue to plant flowers and plants which bees feed on. If there are no flowers within walking distance then it means that your bees will have to fly far for nectar. Even if there are flowers close by, they may not be suitable for bees. By planting in your own yard, you can guarantee that the flowers are the right ones for bees and that the bees are not having to travel to far for the nectar. The further a bee has to travel, the longer to will take them to create honey. That means a smaller yield for them and a smaller yield for you.

Bees love lavender, clover, heather, lilac, thyme, sage and daisies, among others. Do some research into the flowers that bees like which are native to your area.

Join a beekeeping community. This will be your best resource for information on bees and how to care for them. It is great to read up on as much information about bees as you can, but there is some knowledge that can only be gained from having a hive. Take the advice of people who have been raising bees for years.

Watch out for predators. There may already be predators in your area, or more may have moved in after your hive was placed. Bees can be a great source of food for many animals. Be sure to check your hive regularly for any pests which may have moved into your hive and remove them in a responsible manner.

Watch out for cats and other animals roaming your yard. Try to find ways to deter them which are eco-friendly and work. It is not only Winnie the Pooh who loves honey. Bears love nothing more than to come and snack on some fresh, local, organic, honey. If you live in an area which is populated by bears you should ask yourself two questions. Why am I living in an area populated by bears? And what can I do about it?

If a bear gets into your hive, not only will it eat and drink all of your honey, but it will also destroy your hive. This is something which you do not want to happen. The most common way to protect from bears is to install an electric fence to deter them.

Small mammals like to scratch their way into hives to get at the sweet honey inside. You can stop this from happening by building a tall enough stand for your hive. You can also stop very small mammals, such as mice, by making sure your entrance to the hive is small enough. You do not want a mouse taking up home in your hive.

Birds love to eat bees, so do your best not to tempt birds into your yard. If you have a bird feeder hanging close to your hive then you are just asking for trouble. Other bees, and wasps will also try to get into a hive. You can protect from this by keeping your swarm strong. Regular care and checks will keep your bees fighting fit.

Bee hives are also susceptible to other insects moving in. Regular checks of your hive will allow you to see if there are any other being in the hive with your bees. If you find any insects or any sign of illness or disease, you should consult with a local beekeeping community to find the best way to treat it.

Be sure to perform weekly checks on your bees, and in the honey producing months, make sure that you are harvesting enough honey to leave enough space for the bees to make more. Bees will continue to make honey, long after the hive is full. If honey is not taken from the hive then bees will look for larger storage areas, this can mean bees leaving the hive. Be sure to check regularly and empty the frames or add more honey supers to your hive.

The other thing to think about when checking the honey harvest is that the bees have enough honey for food. If you find that the honey production is slow, it may be that they are using most of it as food. If you find that the honey in the hive is showing signs of reduction, it may be time to add more food to the hive. Mix up some more sugar water (one to one ratio) and fill a feeder with it. You can purchase feeders online or you find tutorials on the internet about how to make them.

As winter is setting in, check your hive for good circulation of air and no moisture collecting. If the weather is going to become especially bad, you can wrap your hive in blankets to keep it warmer as the cold sets in. Make sure that the cover fits tightly and there is no damage to it. Check that all the walls are intact and the joints are in good condition.

How to check the hive

Now that we know what to look for, we need to schedule some regular checks. When you first start out, it would be a good idea to check the hive every seven days. As you grow more confident you can make this every ten days. You do not want to check the hive too often as the bees can become angry and it sets back their honey production.

When you are checking the hive, make sure that you have the correct equipment. You should put on your protective clothing before you approach the hive. You should have your smoker (lit) and your hive tool.

Approach the hive and spray some smoke over the entrance to calm the bees who are guarding it. We do not want them to alert the other bees and send them into a panic. Next, open the top cover and spray some smoke under it. Put the cover back down and wait a minute or two for the smoke to take effect. The bees will be nice and calm when you open the cover again.

If at any time during your check you notice that the bees are starting to look at you and notice you, simply spray in a little bit more smoke to calm them again. Now that they are calm, you can remove the outer cover and begin your inspection. Spray in some more smoke if you think that you need to.

Take out the inner cover, with your hive tool if you need to, and scrape any resin or wax from it. Brush any bees from the cover of you need to. Set the inner cover down beside the outer cover.

Lift off the first honey super, using the hive tool if you need to, and set it down. You can place it on a flat surface, one which will not dirty the bottom of the super, or you could place it down on the outer cover to protect it. Do this with any additional honey supers that you have.

You should be down to your deep hive. Spray some more smoke into the hive and wait a minute. Pry off the excluder with your hive tool and place it somewhere safe. Next spray smoke between the frames of the deep super. Remove the first frame and place it in a frame holder (if you have one) or on top of the honey supers. Be careful not to squash any of the bees as you do this.

You may need to use your hive tool to pry out the frames. You will now check the frames one by one and are looking for a few different things:

- How many larvae and eggs are in there.
- The queen.
- Parasites or pests.
- Honey levels.

The larvae will be easy to spot. They will be curled up in the hexagonal structures. A good amount of larvae means that the queen is reproducing and your hive is growing. The eggs are a little harder to spot. They will also be in the hexagonal structures and will look like grains of rice. There should be one egg per hexagon. If there are more than one in a hexagon then other bees may be laying eggs and you should consult with an experienced beekeeper.

If you are finding it difficult to see the eggs, you can hold the frame up and tilt it back and forth. The change in light can make them easier to see. You can also use a magnifying glass.

If you do not see the queen, then look for the eggs. If there are eggs in your hive it means that the queen was there in the last couple of days and there is most likely nothing to worry about. If there are no eggs then it may be time to consult an experienced beekeeper. You may have a pressing question now and that question is likely: how do you spot a queen bee? There are a few things to look for:

- Queen bees are larger than the other bees.
- They have a pointed abdomen.
- Their stinger is smooth and not barbed.
- Her legs are usually splayed apart.
- The other bees will be surrounding her.

Queen bees live for 3-5 years and can be replaced at any time during that period.

When you check the honey levels you may want to harvest some honey or you may want to add some more food to the hive. Once you have performed your check of the deep super, it is time to replace the frames. Make sure that there are no bees in the way as you are replacing each frame. Place them one by one, until all the frames have been replaced. Try to replace each frame in the same order as they were before you took them out.

You can now put the excluder back on the hive. You still have one or two honey supers to inspect. Inspect them in exactly the same way as you did the deep super. Once the honey supers have been replaced, you can replace the inner and outer cover. Be sure during the entire process to be extremely careful and make sure no bees are harmed.

The inspection is done. Once everything is put back together, you can return to a safe distance and remove your protective clothing. Regular checks of your give will ensure that your bees last for a long time.

What Is All The Buzz About Honey?

Your hive is operational and you are taking good care of it. You are ready to reap the rewards for all of your hard work. You will not get a large harvest in your first season as the bees are using all of their energy to build the hive and get it operational. A lot of the honey they produce will be used as food and a lot of the nectar they collect will be consumed too. The population will be low too, it takes time for the population to grow. The more bees you have, the more honey is produced.

July, August and September are your main honey harvesting months. You will begin to see a larger yield from your second year onwards. To be efficient when you are harvesting your honey, you should look for the combs (hexagons) to be 90% full. There is no point in wasting effort with half full combs and depending on which method you use, it could create a lot more work for your bees.

You should also look for 90% of the combs to be capped. When the bees fill the combs with honey, they will cap them to speak the honey inside. Combs which are capped will have a paler, lighter cap on top of them and should be easy to spot. Once you are confident that there is enough honey to harvest, it is time to harvest your honey.

You will use the same equipment that you used when you were making your inspections, but before you get your suit on you should make sure that you have a space to do the honey harvesting. It is preferable to do this indoors as other bees can smell the honey and may come by for a tasty snack.

Once you have your bee suit on, or your veil and protective clothes, you can approach the hive and douse it in some smoke to calm the bees. Then you will use your hive tool, if needed, to pry the cover open and pry out the frames from the honey supers. As you take each frame out, try to move the bees from the frame with a bee brush.

As you take the frames out of the super, place them somewhere where you can cover them with a towel to keep the bees off. When you have all the supers out of the hive, you can transfer them to the place where you will harvest the honey. Make sure to put the hive back together again.

Now that you are in your harvesting area you will need to remove the caps from the honeycomb. To do this you can purchase an electrically heated knife, or you can improvise. You want to use a large flat knife and it does not need to be sharp. Warm your knife and scrape it down each side of the frame at a thirty degree angle. You do not want to scrape off too much of the honey and you do not want to linger or you may burn the honey. If there are any caps remaining, you can scratch them off with a fork, or another device of your choosing. Catch the wax caps in some sort of container to be used later.

Now that you have the caps off, the honey will be on display. There are two ways to get the honey out. You can use a honey extractor or you can use the crush and strain method. Let's talk a little about both.

Extractor

An extractor is a piece of machinery which use centrifugal force to remove the honey from the honeycomb. The main downside of using an extractor is that it costs money. A new honey extractor can cost any-where up to $200-$200US. If you are part of a bee-keeping community you may be able to rent one cheap-ly or may have access to one at no cost.

The extractor works by spinning the frames and using centrifugal force to force the honey from the hon-eycomb. Extractors can come with motors or can be powered by a hand crank. The extractor will typically take three or more frames and it will take the honey from one side at a time.

Extractors are quick and do not make much mess. Once the honey has been forced out, it will run down the sides of the machine and into the bottom. When you have enough honey in the bottom of the ex-tractor you can open the tap and let the honey run through a strainer to take out any imperfections.

One of the main advantages of using an extractor is that the honeycomb remains intact which means that the bees do not need to rebuild it. If the bees are not working on rebuilding their home, then they have more time to make honey for you.

Crush & Strain

Once you have the caps off your honey you can crush and strain the honeycomb. To do this you simply scrape the honeycomb from your frame and into a large container. Once you have everything in the container, you crush the honeycomb and honey as finely as you can. When it is crushed you pour the mixture through a strainer to get rid of the pieces of honeycomb.

This method is cheaper than using an extractor as there is no special equipment to buy and it can be quick, though the straining can sometimes take a long time and you may need to leave it running through a

strainer overnight to get all of the honey. The main disadvantage of this is that the honeycomb is destroyed.
The bees will have to work on replacing it before filling
it with more honey for you.

Canning

Glass jars are the best way to store your honey.
Buy some mason jars and sterilize them. Once you
have done that, if the honey is liquid enough, you can
pour the honey straight into the jars and put the lids on.
Honey can be stored on a shelf in your kitchen, out of
direct sunlight and can last indefinitely.
If there are any imperfections still in the honey,
they will rise to the top after a day or two and you can
skim them off. Honey may crystallize over time, but you
can return it to its original state by placing the jar in
some boring water. This will make the honey liquid
again. You may also notice the honey darken over time
and change taste slightly. This is also perfectly normal.
Now the best part, you get to devour and enjoy
your honey. Make sure to replace the frames in the hive
so that the bees can get back to work. Those lazy bees!

Beeswax

The beeswax that you collected earlier can also
be used in a variety of ways. Take what you collected
and melt it in a pot. Once you have done that, pour it
into a container and wait for it to separate. Any honey
that was scraped off with the wax will be on the bottom
and the wax will rise to the top. When it has hardened,

you can separate the two and do some wonderful things with the wax:
- Make candles
- Lip balm
- As a sealer
- Moisturizer
- In certain types of cooking
- Pain relief
- Lubrication
- And more

Honeycomb

The honeycomb can also be kept and eaten. You may want to take chunks if you are using the extractor, or you may want to take the leftover bits after it has been strained. It can be eaten as it is or you can add it to a variety of dishes. Honeycomb is great sprinkled over a salad to add a sweetness to it. It can be stirred into yoghurt or added to a bowl of oatmeal. There are an endless amount of ways that it can be used. It can also be stored in the same way as honey, in a jar and at room temperature.

Time To Buzz Off

And that is pretty much that. There are so many things that you can do with honey that i is beyond the scope of this book. Not only is it delicious, but it can be used to treat pain, infections, allergies and so much more. The honey you harvest in your own yard will have vastly more nutritional value and healing proper-ties than that which you buy in a store.

There is only one more thing to do now and that is to go get a hive and start your honey production. As you can see, it is not all that hard. If you put in a little time and effort, you can harvest a lot of honey and help to pollinate the surrounding area. In any given year you should expect between 20-250lbs of honey. That is amazing!

Did I say there was only one more thing to do? Sorry, I should have said two. You need to share the honey with your friends. You are going to have so much honey, and it is going to taste so good, that your friends are going to be begging you for it.

Thank you for taking an interest in beekeeping, not only are you starting down a path towards honey harvesting, but you are also helping to save the world and the bee population. I wish you the best of luck on your journey.

www.ingramcontent.com/pod-product-compliance
Lightning Source LLC
Chambersburg PA
CBHW020714180526
45163CB00008B/3081